DEADLIEST ANIMALS

# BLACK MAMBA

A Crabtree Branches Book

Amy Culliford

I0792580

Crabtree Publishing
crabtreebooks.com

# School-to-Home Support for Caregivers and Teachers

This high-interest book is designed to motivate striving students with engaging topics while building fluency, vocabulary, and an interest in reading. Here are a few questions and activities to help the reader build upon his or her comprehension skills.

## Before Reading:

- *What do I think this book is about?*
- *What do I know about this topic?*
- *What do I want to learn about this topic?*
- *Why am I reading this book?*

## During Reading:

- *I wonder why...*
- *I'm curious to know...*
- *How is this like something I already know?*
- *What have I learned so far?*

## After Reading:

- *What was the author trying to teach me?*
- *What are some details?*
- *How did the photographs and captions help me understand more?*
- *Read the book again and look for the vocabulary words.*
- *What questions do I still have?*

## Extension Activities:

- *What was your favorite part of the book? Write a paragraph on it.*
- *Draw a picture of your favorite thing you learned from the book.*

# TABLE OF CONTENTS

# A DEADLY SNAKE

Africa is a **vast** land full of deadly animals. Hippos, elephants, crocodiles, and lions probably come to mind. After all, they are large and definitely dangerous. Perhaps another dangerous animal deserves more attention—the black mamba. The black mamba is one of Africa's deadliest snakes.

Thousands of people die every year in Africa from **venomous** snake bites. Without immediate medical attention, a black mamba bite will often result in death.

Black mamba snakes are feared in Africa. Their bite is deadly and they are known to be aggressive when threatened. It doesn't help that they are one of the fastest snakes in the world. A black mamba that is cornered, or feels threatened, is a dangerous animal.

BLACK MAMBAS CAN MOVE MORE THAN 12 MILES PER HOUR (20 KPH). THE SIDEWINDER RATTLESNAKE, HOWEVER, IS ACTUALLY THE FASTEST SNAKE IN THE WORLD. IT MOVES 18 MILES PER HOUR (29 KPH).

sidewinder rattlesnake

# WARM AND DRY

Black mambas live in many African countries that lie south of the Sahara Desert. They prefer dry environments. They make their home in rocky areas along rivers and in wooded grasslands.

termite mound

Black mambas mostly travel over ground, but sometimes they make their way up trees. They hunt by day and sleep at night in hollow trees or empty termite mounds.

**BLACK MAMBAS ARE COLD-BLOODED. THEIR BODY TEMPERATURE DEPENDS ON THE TEMPERATURE OF THEIR SURROUNDINGS.**

# FAMILY LIFE

Saying black mambas have a family life may be a stretch. But, like all animals, mambas must **reproduce** to survive. Courtship begins when males perform a dance-like combat with other males. The winner of the battle gains mating rights with the female.

Male mambas wrestle for mating rights.

After mating, the male and female snakes do not interact again. About 60 days later, the female lays eggs in a tree hollow or other safe place, then leaves never to return—that's not much of a family life!

Baby mambas look like tiny adult mambas.

# BLACK MAMBAS

Black mambas are not black. They are typically brownish-gray on top and grayish-white underneath. They get their name from the black color inside their mouth.

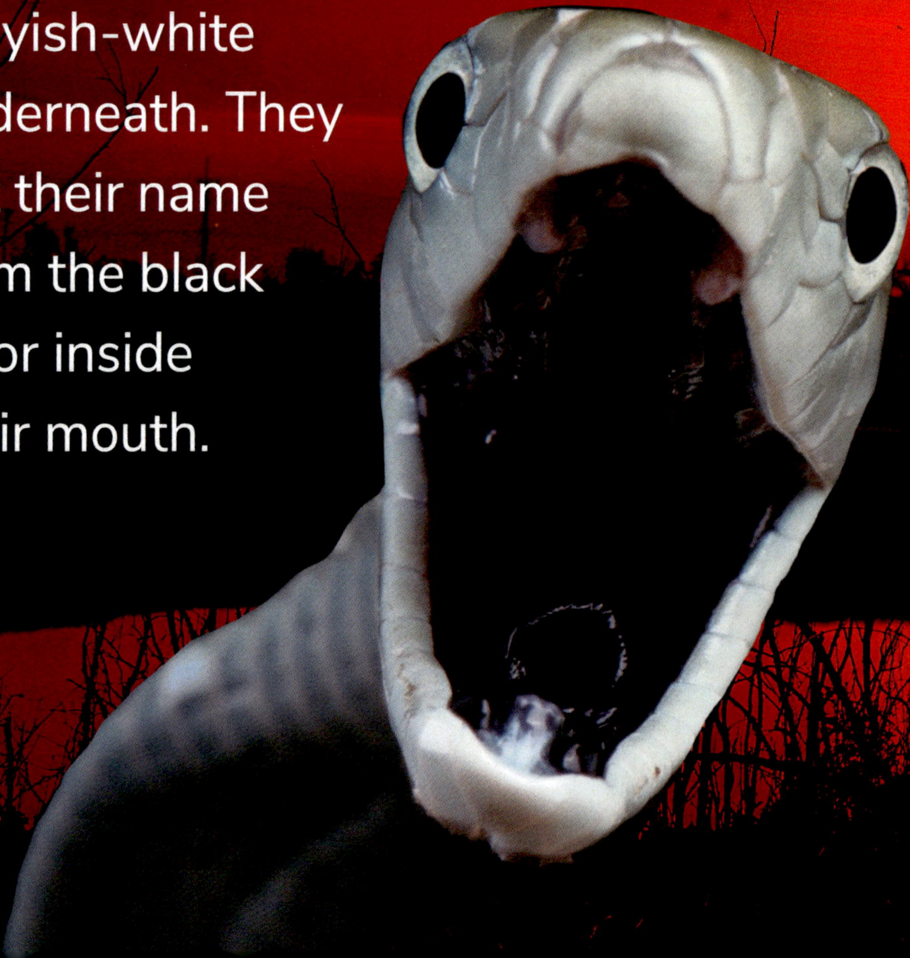

BLACK MAMBAS ARE THE LONGEST VENOMOUS SNAKE IN AFRICA, REACHING 14 FEET (4.3 METERS). THEY ARE THE SECOND LONGEST VENOMOUS SNAKE IN THE WORLD. ONLY ASIA'S KING COBRA IS LONGER.

# MAMBA SNAKES

Black mambas are one of four **species** of mamba snakes. All four are very fast and very venomous. The black mamba is a ground dwelling snake, while the others live in trees.

**JAMESON'S MAMBA:**

THIS SPECIES LIVES IN HOT CENTRAL AFRICA.

**EASTERN GREEN MAMBA:**

THIS SPECIES LIVES IN THE COASTAL REGIONS OF EAST AFRICA.

**WESTERN GREEN MAMBA:**

THIS GREENISH-YELLOW SNAKE LIVES IN THE COASTAL RAINFOREST OF WEST AFRICA.

DID YOU KNOW? THE NUMBER OF ROWS OF SCALES HELPS IDENTIFY A SNAKE SPECIES. BLACK MAMBAS HAVE BETWEEN 248 AND 281 SCALES ON THEIR BELLY. THESE SCALES HELP THEM MOVE OVER FLAT SURFACES.

# DEADLY WEAPONS

## WEAPON 1—FANGS

Black mambas have two fangs at the front of their upper jaw. Adult fangs are about 1/4 inch (6.5 mm) long. Their purpose is to inject venom into **prey**.

## Weapon 2—Venom

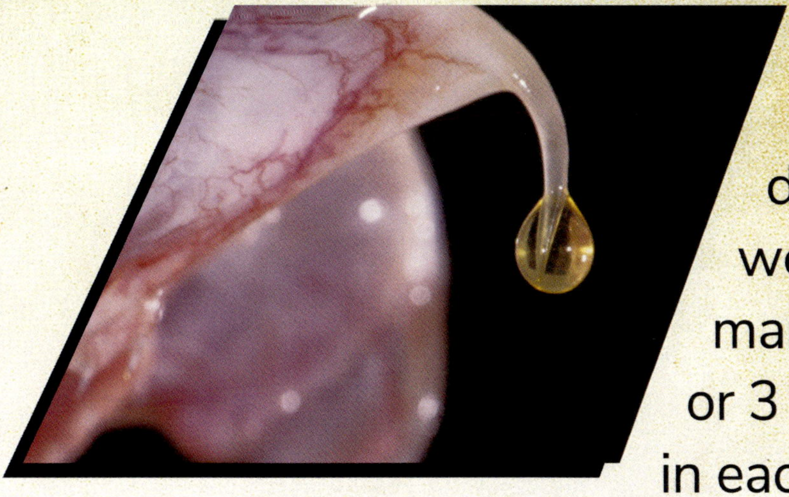

Venom is the mamba's deadliest weapon. Young mambas have 2 or 3 venom drops in each fang. Adults can have 20 drops per fang. It takes only 2 drops to kill an adult human.

## Weapon 3—Speed

Speed is important to the black mambas' survival. They not only move fast, but they can also strike in a fraction of a second. Black mamba venom is also fast. It can **paralyze** or kill prey in minutes.

# Weapon 4—Color

The black mamba's brownish-gray coloring is **camouflage**. It helps the snake blend in with mud and rocky hillsides.

A black mamba often warns enemies before attacking or defending itself. It raises its head, and up to a third of its body length, off the ground. It flattens its neck into a hood, and opens its black mouth. Sometimes it hisses at its enemy.

# WEAPON 5—KEEN SENSES

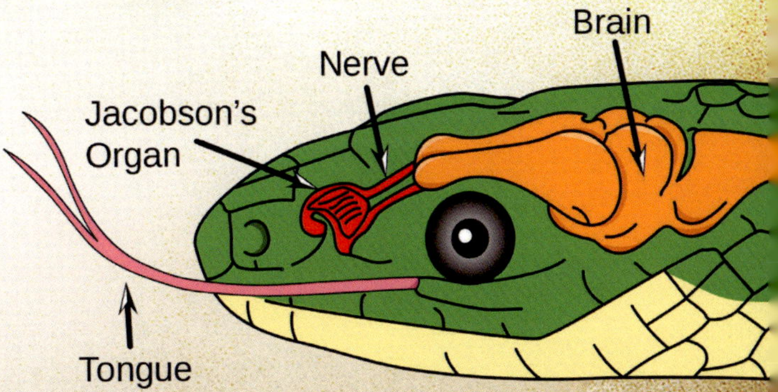


Black mambas have a **Jacobson's organ** in their snout. The organ is a patch of **sensory** cells. When the mamba flicks its tongue in and out, the cells help the snake detect odors and moisture in the air.

# SLITHERING CARNIVORES

Like all snakes, black mambas are carnivores—they eat other animals to survive. Black mambas prey on rodents, birds, and other snakes. They have good eyesight and can sense movement. However, the black mamba hunts by sense of smell, using its Jacobson's organ to track prey.

Black mambas can live about 11 years in the wild.

# IN FOR THE KILL

A black mamba will sometimes lie in wait for prey. When prey gets close, the mamba strikes. Other times, the snake uses its famous speed. It will chase down prey, and attack in the blink of an eye.

After injecting venom into a victim, the black mamba retreats and waits for the venom to do its deadly work. Retreating protects the mamba from struggling prey.

**DID YOU KNOW? VENOMOUS DESCRIBES A DANGEROUS FLUID THAT IS INJECTED. POISONOUS DESCRIBES A DANGEROUS SUBSTANCE THAT IS INGESTED.**

Once bitten, its victim has no chance. Mamba venom works quickly. It takes a tiny amount of black mamba venom to kill. The venom shuts down the animal's nervous system. This causes paralysis and death. When its victim no longer moves, the black mamba moves in and swallows its meal whole.

A black mamba jaw opens in two directions, allowing it to swallow prey much bigger than its skull.

# BLACK MAMBAS AND PEOPLE

Black mambas prefer to live far away from people. However, **habitat** destruction sometimes forces them to live close to humans.

Despite their reputation for being aggressive, black mambas are normally shy. If given the chance, they will almost always flee rather than attack.

**ANTIVENOM IS A TREATMENT FOR VENOMOUS BITES AND STINGS. IT CAN STOP FURTHER DAMAGE, BUT CANNOT REVERSE DAMAGE THAT HAS ALREADY BEEN DONE.**

# MAMBA ATTACK!

In 2008, a young man was training to be a safari guide at a South African wildlife college. A black mamba was spotted where students often walked. The trainee and staff members caught the snake to remove the danger. As the young man helped put the snake in a jar, he told staff members he thought the snake brushed against his finger. Within an hour, the trainee's vision blurred. He collapsed and died a short time later. Doctors concluded the trainee died from black mamba venom. The young man didn't even realize he had been bitten!

BLACK MAMBAS ARE LARGE, FAST, AND VENOMOUS. THAT'S WHY THEY ARE CONSIDERED ONE OF THE WORLD'S DEADLIEST ANIMALS.

# GLOSSARY

**camouflage** (KAM-uh-flahzh): Coloring that makes animals look like their surroundings

**cold-blooded** (KOHLD BLUHD-id): Animals that have body temperatures that change with the temperature of their surroundings

**habitat** (HAB-i-tat): The natural place where a plant or animal lives

**Jacobson's organ** (JAY-kub-sunz OR-gun): A body part located on the roof of many animals' mouths that helps them smell things

**paralyze** (pair-a-LYZE): To make unable to move

**prey** (PRAY): An animal that is hunted by another animal for food

**reproduce** (ree-pruh-DOOS): To produce offspring

**sensory** (SEN-suh-ree): Having to do with the senses of smell, hearing, sight, taste, and touch

**species** (SPEE-sheez): A certain type of animal or plant

**vast** (VAST): Huge area

**venomous** (VEN-uh-muhs): Animals that have poisonous fluid that they deliver by injection

# Index



# Websites to Visit

https://kids.britannica.com/kids/article/mamba/543521

www.krugerpark.co.za/krugerpark-times-17-facts-about-the-black-mamba.html

https://kids.kiddle.co/Black_mamba

# About the Author

## Amy Culliford

Amy Culliford holds a Bachelor of Fine Arts. She has worked as a classroom drama teacher and led after-school drama programs. She avoids snakes of any kind.

The author would like to thank David and Patricia Armentrout for their research and help on this project.

## Crabtree Publishing

crabtreebooks.com 800-387-7650



**Produced by: Blue Door Education for Crabtree Publishing**
**Written by: Amy Culliford**
**Designed by: Jennifer Dydyk**
**Edited by: Tracy Nelson Maurer**
**Proofreader: Crystal Sikkens**

| | |
|---|---|
| Hardcover | 978-1-4271-5411-8 |
| Paperback | 978-1-4271-5417-0 |
| Ebook (pdf) | 978-1-4271-5423-1 |
| Epub | 978-1-4271-5429-3 |
| Read-along | 978-1-4271-5435-4 |
| Audio book | 978-1-4271-5441-5 |

Printed in the U.S.A./072025/CP20250721

**Published in Canada**
**Crabtree Publishing**
616 Welland Avenue
St. Catharines, Ontario
L2M 5V6

**Published in the United States**
**Crabtree Publishing**
347 Fifth Avenue
Suite 1402-145
New York, NY 10016

Photographs: Cover photo © Robin Winkelman | Dreamstime.com, graphic splat on cover and throughout © Andrii Symonenko /Shutterstock.com, Page 4 © Susan Schmitz/Shutterstock.com,page 5 map © maodoltee/Shutterstock.com, mamba page 5, 6, 8 © NickEvansKZN/Shutterstock.com, page 7 © Fine Art Photos/ Shutterstock.com, pages 6-7 background image © Giordano Aita/Shutterstock.com, page 8 globe © MarcelClemens/Shutterstock.com, page 9 (top) © Vadim Petrakov/Shutterstock.com, (bottom) © Tallies/Shutterstock.com, page 10 © © Gerhardt Nieuwoudt https://creativecommons.org/licenses/by-sa/1.0/deed.en, Page 11 (top) © Dylan leonard/Shutterstock.com, (bottom) © NickEvansKZN/ Shutterstock.com, pages 12-13 background photo © sarintra chimphoolsuk/ Shutterstock.com, page 12 © reptiles4all/Shutterstock.com, page 13 (top) © NickEvansKZN/Shutterstock.com, (bottom) © Pieter Oosthuizen/Shutterstock.com, pages 14-15 background photo © Manamana, page 14 © Jiri Prochazka/ Shutterstock.com, page 15 (top) © himanshu_alwaria/Shutterstock.com, (bottom) © J.A. Dunbar, Page 16 © Jorgen Mus/Shutterstock.com, page 17 © Joe McDonald/Shutterstock.com, page 18 © Cormac Price/Shutterstock.com, page 19 (top) © Stu Porter/Shutterstock.com, (bottom) © © Fred the Oyster (Wikipedia) https://creativecommons.org/licenses/by-sa/4.0/, pages 20-21 background photo © Manamana, page 20 © Reptiles4all/Shutterstock.com, page 21 (top) © NickEvansKZN/Shutterstock.com, (bottom) © Tobie Oosthuizen/Shutterstock.com, page 22 © Tad Arensmeier (Wikipedia) https://creativecommons.org/ licenses/by-sa/3.0/deed.en, page 23 (top) © NickEvansKZN/Shutterstock.com, (bottom illustration) © Martial Red/Shutterstock.com, page 24 © reptiles4all/ Shutterstock.com, page 25 © Karl van der Westhuizen/Shutterstock.com, page 26 © Karl O'Neill /Shutterstock.com, page 27 (top) © Andrea Izzotti/Shutterstock.com, (bottom) squeezing snake venom © MemoryMan/Shutterstock.com, vials © SashaMagic/Shutterstock.com, page 28 © Stu Porter/Shutterstock.com, page 29 © NickEvansKZN/Shutterstock.com

**Library and Archives Canada Cataloguing in Publication**

Title: Black mamba / Amy Culliford.
Names: Culliford, Amy, 1992- author.
Description: Series statement: Deadliest animals | "A Crabtree branches book". | Includes index.
Identifiers: Canadiana (print) 20210218851 | Canadiana (ebook) 2021021886X | ISBN 9781427154118 (hardcover) | ISBN 9781427154170 (softcover) | ISBN 9781427154231 (HTML) | ISBN 9781427154293 (EPUB) | ISBN 9781427154354 (read-along ebook)
Subjects: LCSH: Black mamba—Juvenile literature.
Classification: LCC QL666.064 C85 2022 | DDC j597.96/4—dc23

**Library of Congress Cataloging-in-Publication Data**

Names: Culliford, Amy, 1992- author.
Title: Black mamba / Amy Culliford.
Description: New York : Crabtree Publishing, 2022. | Series: Deadliest animals - a Crabtree branches book | Includes index.
Identifiers: LCCN 2021021925 (print) | LCCN 2021021926 (ebook) | ISBN 9781427154118 (hardcover) | ISBN 9781427154170 (paperback) | ISBN 9781427154231 (ebook) | ISBN 9781427154293 (epub) | ISBN 9781427154354
Subjects: LCSH: Black mamba--Juvenile literature. | Dangerous animals--Juvenile literature.
Classification: LCC QL666.064 C85 2022 (print) | LCC QL666.064 (ebook) | DDC 597.96/42--dc23
LC record available at https://lccn.loc.gov/2021021925
LC ebook record available at https://lccn.loc.gov/2021021926